AMAZON ECHO DOT IN 1 HOUR

Amazon Echo Dot in 1 Hour:
The Complete Guide for Beginners - Change Your Life, Create Your Smart Home and Do Anything with Alexa!

Joel Goodwin

Joel Goodwin
2017

Copyright © 2017 by Joel Goodwin

All rights Reserved. No part of this publication or the information in it may be quoted from or reproduced in any form by means such as printing, scanning, photocopying or otherwise without prior written permission of the copyright holder.

Disclaimer and Terms of Use: Effort has been made to ensure that the information in this book is accurate and complete, however, the author and the publisher do not warrant the accuracy of the information, text and graphics contained within the book due to the rapidly changing nature of science, research, known and unknown facts and internet. The Author and the publisher do not hold any responsibility for errors, omissions or contrary interpretation of the subject matter herein. This book is presented solely for motivational and informational purposes only.

First Printing: 2017

ISBN 978-1544030586

Contents

Introduction 1

Chapter 1: Getting to Know the Secrets of Your Dot 5
   Finally, Familiarizing Yourself with the Echo Products. 5
   The Amazon Echo 6
   The Echo Dot 9
   The difference between the first and second generation Dot 10
   The Echo Spatial Recognition Made its Debut 12
   How Alexa Works 15

Chapter 2: The Primary Features of Your Echo Dot 18
   Let the Music Go Wild 18
   Radios are back 24
   Audiobooks Never Felt More Real 25
   Traffic Updates Right at Your Finger tips 27
   All the Weather News 28
   Sports Update 29
   Allowing Alexa to Wake You Up 30
   The Short Cat Nap Burst 33
   Creating a Bridge Between Google Calendar and Alexa 36

Chapter 3: The Advanced Secrets of Echo Dot 38
   Dot as Your Bluetooth Speaker 38
   Creating Your Very Own List 39
   A Bliss for Big Families 41
   Changing the Name of Your Dot 42
   Delete All of Your Recordings 43
   Some Fun Games to Jump Into! 44
   Finding the Location of Your Package 45
   Your very own kitchen assistant in the form of Echo Dot 45
   Using Echo Dot as your personal calculator 46
   The Power of Dot in the Form of Skills 46

Chapter 4: Automating Your Home 52
   The secret behind everything 52
   The list of compatible devices 52

An alternative way to add your devices 59
The Final Step 60

Chapter 5: A Selection of Troubleshooting Tips and FAQs 62

Lost Wi-Fi connection 62
Unable to Find Connected Home Devices.........................63
Hard-resetting Your Dot 63
Dot not being able to hear properly....................................64
F.A.Q Section 66

Conclusion 69

Joel Goodwin

Introduction

When it comes to companies and organizations that lead the cradle of innovation, very few come close to the mantle created by giants such as Apple, Google or even Amazon!

That's right, I'm talking about the Amazon we all come have come to know and love. Surprised? Don't be! When Amazon was first established during the early '90s, nobody expected it would turn into such a corporate mega mammoth. The only success they found in the past was as an online book store. However, soon after they realized they were ready to expand, they started to diversify their reach and make even more products available, which included DVDs, CDs, MP3s, video games, electronics and so on! Moving a little bit forward, various other services such as audio, video and even book streaming were added in their cluster as well.

But did Amazon stop there? Of course not! Soon after, they claimed their throne of being the largest online retailer shop in America! They started to manufacture amazing gadgets such as the Kindle, Fire TV, and the Fire Tablet.

But that's not what you are here for, right?

You are here to know about their latest and greatest creation that took the entire industry by storm!

What is the Echo, you want to know? Well, these are the products that brought consumer grade "futuristic" devices to the palms of normal people.

Back in 2011 Amazon clearly understood that the whole market and common individuals were slowly shifting toward turning their whole household into futuristic masterpieces through the usage of so called "smart" or "automated" gadgetry. Products that gave users an opportunity to acquire even a limited amount of automated productivity were selling like hot cakes straight out of the oven, despite having a high price tag!

Having a firm understanding of that concept, Amazon moved forward by taking a bold move and attempting to penetrate the market for smart home devices back in 2014. They did this by releasing their first ever Amazon Echo in hopes that even the common people (not billionaires) would be able to jump on the band wagon and turn their home into a futuristic marvel without dishing out loads of cash.

And unsurprisingly, I should say, the Amazon Echo became an instant hit and won the hearts of millions of people out there.

Flash forward to the year 2016, when Amazon released their latest iteration of the Echo product line-up and blessed the world with the most affordable and perhaps adorably cute Echo device yet: the Echo Dot.

Although these devices were originally advertised as being Bluetooth speakers, they were far from it and they boasted a heap of different functionalities as well.

All of the above mentioned features were only made possible thanks to the inclusion of their most sophisticated and highly skilled artificially intelligent software, codenamed "Alexa."

Alexa was the game changing and revolutionizing software installed in all Amazon's Echo products that seamlessly allowed user interactivity between the device and the human using just the voice. Any command and gesture could be issued just by saying it.

And that scheme was further pushed forward with the inclusion of over 1500 different apps which came in the form of "Skills." Any form of action that could be translated into audio could be accomplished by the Echo Dot now!

When the concept and prototype of Alexa was first released into the public, it gained nothing short of extremely positive responses where upon, during its launch, its reception was compared to that of the iPhone. You can imagine how big a deal that was!

With the power of Alexa, nothing was out your grasp anymore. Want to hear the latest music? Just shout out the name! Need the latest weather forecast or even the latest news? Alexa will bring it to you in a moment's notice.

Alexa also allowed the Dot to fully interact with all of the compatible smart devices around your house. You could even control of the room's temperature just by using your voice.

All of these functionalities and further innovations led to the creation of the fantastic and ever-evolving Echo "Dot." The latest, and perhaps the most affordable, device in the whole Amazon Echo series that gave you total control over all the smart devices in your house, while looking incredibly cute and stylish.

If that's not futuristic, then I don't know what is! While the bundled up user manual that came with your Dot was definitely helpful, Amazon held a whole lot of the Dot's true potential back.

And that was the sole purpose behind the idea of this book. This guide is not only designed to become the next simple instruction booklet, but it's also the most ultimate and definite guide out there to help you learn the tricks and trades of your Echo Dot.

So, let's stop wasting more time and jump right into unravelling the secrets of your Dot!

Joel Goodwin

Chapter 1: Getting to Know the Secrets of Your Dot

After reading that hefty introduction, I am pretty sure you are ready to jump right into the world of the Amazon Echo and explore the secrets of your ever so intelligent Dot. But before I start to inject you with all the knowledge out there, it is essential that you know a little bit about all the other products in the Echo line.

Yes, you read that right; your Dot isn't the only Echo, as it has a number of different siblings out there, which altogether comprises the Echo line of products.

Finally, Familiarizing Yourself with the Echo Products

Back in 2014, when Amazon first released the Amazon Echo, even they didn't imagine that it would turn into such a massive hit! With the aim of bringing the future of home automation within the grasp of the common people, Amazon took a huge leap of faith when it announced the first Alexa compatible intelligent blue tooth speaker, the Amazon Echo, which stood as their first product in the soon-to-be hit series.

Unsurprisingly, though, thanks to the myriad of features the device boasted, it soon turned into a superlative hit amongst friends and families, catapulting itself into the position of being the ultimate holiday gift, which is probably why you got the Dot yourself this holiday season, right?

Building upon the success of the first Echo, Amazon made it their motto to innovate upon what they had already created, and bring it closer to perfection, while making sure that the device remains within the reach of everyday citizens.

In that essence, as of March 2017, the Echo line of products is comprised of the following available hardware:

- The Amazon Echo
- The Amazon Tap
- The Amazon Echo Dot

Even though the core functionalities of the devices intersect each other, each of them sports something special that makes them what they are.

Since the focus of this book is primarily going to be the Echo Dot, I will briefly run you through its predecessors and elaborate the core difference between the Echo Dot and the other devices.

The Amazon Echo

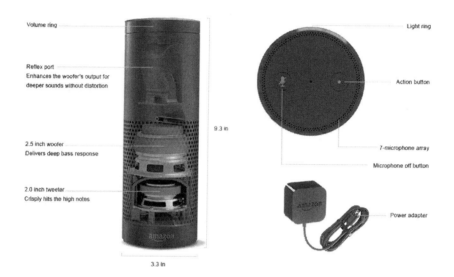

At its heart, Amazon Echo is an elegantly designed "Smart" Bluetooth speaker developed by Amazon. It has a cylindrical shape with a height of approximately 9.25 inches. To someone who is unaware of its capabilities, it might seem like an ordinary speaker, but anyone who would spend a few minutes with it will begin to realize that this device has much more than meets the eye.

The "smart" part of this device comes from the built-in personal assistance, which has been named Alexa. If you are a fan of the wise-cracking, fictional, artificially intelligent character Jarvis from the Iron Man movies, then you are going to absolutely adore Alexa!

You will be able to manipulate all the features of your speaker by commanding Alexa only using your voice. These features can range from playing your favorite song, adding items to your shopping list,

Joel Goodwin

getting instantaneous traffic updates or even turning your lights on and off! We will be discussing all of these and more throughout the course of this book, so hold onto your handles and prepare to be blown away.

Boasting a height of 9.25 inches, the Echo was the first step towards creating nothing short of a revolutionary product.

Known as "Project D" during its development, the Echo introduced the almost sentient and perhaps one of the most advanced AI known as "Alexa," which would later become a staple companion for all future products in this line.

When compared to the Echo Dot, the prime difference here has to be size. Echo Dot takes the mantle for being the most portable device of the three, but the speakers on Echo are relatively superior when compared to those of the Dot. The technical specifications of the device are as follows:

Size	9.3" x 3.3" x 3.3" (235 mm x 84 mm x 84 mm)
Weight	37.5 oz. (1064 grams) *Actual size and weight may vary by manufacturing process*
Wi-Fi Connectivity	Dual-band, dual-antenna Wi-Fi (MIMO) for faster streaming and fewer dropped connections than standard Wi-Fi. Supports 802.11a/b/g/n Wi-Fi networks. Does not support connecting to ad-hoc (or peer-to-peer) Wi-Fi networks.
Bluetooth Connectivity	Advanced Audio Distribution Profile (A2DP) support for audio streaming from your mobile device to Amazon Echo and Audio/Video Remote Control Profile (AVRCP) for voice control of connected mobile devices. Hands-free voice control is not supported for Mac OS X

	devices.
Audio	2.5 inches woofer and 2.0 inches tweeter
System Requirements	Amazon Echo comes ready to connect to your Wi-Fi. The Alexa App is compatible with Fire OS, Android, and iOS devices and also accessible via your web browser.
Warranty and Service	1-year limited warranty and service included. Optional 1-year, 2-year, and 3-year extended warranty available for U.S. customers sold separately. Use of Amazon Echo is subject to the terms found
Included in the Box	Amazon Echo, power adapter/cable (6 ft.), and quick start guide

At the time of writing, the Echo has a price tag of **$180**.

Joel Goodwin

The Echo Dot

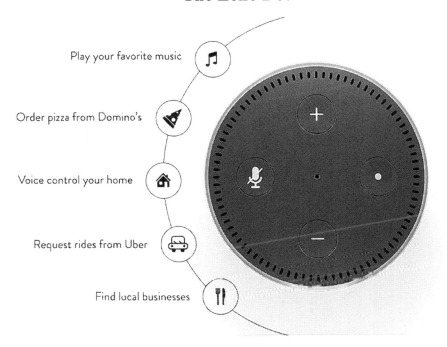

When Amazon released the Echo Dot in May 2016, their biggest goal was to achieve the same level of flexibility and functionality offered by the bigger Echo or Tap, and pare it down into a much more portable and affordable package! With the Echo, they really kept that promise and delivered even more!

At the size of a hockey puck, boasting a dimension 1.5 x 3.3 x 3.3 inches and weighing only 0.6 pounds, the Amazon Echo Dot redefined the meaning of the phrase, "Big things come in small packages!"

And this package came at only a mere **$50! Or, at least, the 2nd Gen version did. The first one costs about $89.99 at the time of this writing.**

When the Dot was released earlier this year, it quickly became Amazon's hottest selling product and ran out of stock like freshly baked hot cakes!

In terms of functionality, Amazon made sure that no sacrifices had to be made during the transition to a smaller form factor, and as such, the Dot is basically able to pull off everything its elder sister Echo is capable of!

This is just the bigger Amazon Echo, packed up in a nice little hockey puck and without a doubt the perfect product for anyone hoping to enter the world of home automation without burning a hole in their pocket.

Now that you are at the end of the section, I'm presuming you're looking for the technical specs?

Well, here's the thing! Due to its immense popularity, right after the Dot was released into the mass market, Amazon took a step beyond and went further, to immediately improve upon the slight limitations of the Echo Dot. And thus, the Echo Dot 2, or Dot 2nd gen, was born.

The difference between the first and second generation Dot

Keep in mind that both the devices, along with all the other devices in the Echo line, are powered by Alexa. Therefore, you will be getting most, if not all, of the features of Alexa simultaneously across all platforms.

However, the difference between each iteration of devices is mostly from the perspective of either hardware or aesthetics.

The same applies for the 2nd Generation Dot.

Joel Goodwin

First, let's talk about the aesthetics. The 2nd Gen Dot is, hands down, the prettiest device of the whole Echo line.

While both the devices feature the same puck design and have similar dimensions, the 2nd Gen Dot has a much glossier and smoother finish. One downside, at least in my opinion, would be the fact that the device sacrifices the rotating volume controlling disk of the first generation in favor for a more traditional button mechanism to keep the costs down.

It is also slightly more compact and lighter in comparison to the first Dot and is not pre-packaged with a 3.5mm cable.

Keep in mind that the 2nd Gen Dot doesn't just come in a pearl color, but it also sports a black color as well.

Now for the hardware. Mainly two aspects from the original device were improved upon with the latest iteration.

Superb Voice Recognition

While the Echo Dot had a pretty solid and all-rounder voice recognition system, there were still some complaints on the system being a little bit shoddy to say the least in comparison to its elder brothers, the Echo and Amazon Tap.

Amazon took this complaint seriously and charged right ahead to improve the already installed "seven microphones" of the Echo with a new and improved speech processor. This does not only help the Far Field Voice recognition mechanism to work better than ever, but it also helps enhance the Dot's accuracy in understanding and issuing the voices.

The Echo Spatial Recognition Made its Debut

This is arguably one of the more in demand features, and Amazon did a splendid job in introducing this alongside the 2nd generation Dot.

The Echo Spatial Perception system, or ESP for short, allows the user to make sure that if you have multiple Echo devices in your house, only one of them (the one nearest to you) receives the instructions, instead of all the devices at once.

This also helps the device to understand your voice more clearly and prevents comedic situations from arising.

Just imagine how you would feel if you told Alexa to order pizza and all three Echo devices start to dial Dominos to order pizza!

Creepy, right? Well, that won't happen ever again.

And the best part about this feature is that, while the ESP was introduced in the Dot 2, it will still be available to all the previous products thanks to a future software update.

Below is a breakdown of the specifications of an Echo 1 vs. an Echo 2 for your convenience.

	Amazon Echo Dot 1st Gen	Amazon Echo Dot 2nd Gen
Dimensions	1.5 x 3.3 x 3.3 inches	1.3 x 3.3 x 3.3 inches
Weight	250 grams	163 grams

	Amazon Echo Dot 1st Gen	**Amazon Echo Dot 2nd Gen**
Buttons	Mute, Action	Mute, Action, Volume
Lights	Light Ring	Light Ring
WiFi	802.11a/b/g/n Dual Band (2.4 GHz and 5 GHz)	802.11a/b/g/n Dual Band (2.4 GHz and 5 GHz)
Bluetooth Audio Input	Yes	Yes
Bluetooth Audio Output	Yes	Yes
AUX Audio Input	No	No
AUX Audio Output	Yes	Yes
Speaker	Small Speaker only for Alexa voice feedback	Small Speaker only for Alexa voice feedback
Power	Adapter (not portable)	Adapter (not portable)

Amazon Echo Dot in 1 Hour

	Amazon Echo Dot 1st Gen	**Amazon Echo Dot 2nd Gen**
Alexa Activation	Voice or Action button	Voice or Action button
All Alexa Features	Yes	Yes
Speech Processor	No	Yes
ESP	Coming via OTA update	Yes
Price	89.99	49.99
Warranty	1 Year	90-day
Availability	Only through Alexa devices	Amazon website
Colors	Black	Black, White

And with that you are now fully introduced to the world of the Amazon Echo as well as the Dot!

Let's dig a little bit deeper and explain what exactly Alexa is.

Joel Goodwin

How Alexa Works

If you have ever watched any science fiction movie or read a story of that genre, you mostly definitely are familiar with the concept of artificial intelligence.

Artificial intelligence is a programmed entity that has a consciousness of its own and is often feared to someday rise up and take over the world (think of Skynet from *Terminator*).

A vague equivalent to that in our world would be Apple's Siri or Microsoft's Cortana. Both of them are designed to become digital companions who will not only help you navigate through your device, but will also try to keep you occupied during the times of your need by offering a plethora of different services.

Alexa does the same thing for the Echo line of products, but she is much more advanced and complex when it comes to its functionalities. Plus, just like a human brain, Alexa has been programmed in such a way that it is constantly learning and evolving to make sure it can give you the best possible experience, while at the same time tries to understand you and your wants even better.

This is the brain of the Echo products, and it is stored in a cloud of its own. The more you interact with her, the smarter she gets by slowly molding herself around your speech patterns, likes and dislikes, accent and vocabulary.

Alexa forms a kind of a vocal bridge between you and the Dot, which allows you to interact with the device and pass on your commands with the power of your voice.

With Alexa, your Dot has the power to become the controlling mother hub for all your smart devices scattered throughout your home. Thermostat, lighting, power switches; anything that had the word "smart" accompanying its name, Dot can bless you with the power of controlling it just by using your voice!

But that's not even the tip of the iceberg. As you will soon see, Alexa is one of the most sophisticated and versatile artificially intelligent companions on the planet!

Throughout the whole book, I will be discussing each one of Dot's features and explain it to you in an easy to follow manner, so you can become the master of your new Dot in no time!

While there are definitely some advanced users out there who already know how to set up the Dot, this book is both dedicated to them, as well as the individuals who have just entered the amazing world of the Echo!

For the absolute beginners out there, you should not worry at all, as setting up the device is as easy as saying one, two, three! But to make the process even more accessible and simplified, here I have included a step-by-step outline of what you should do in order to set up your device right from the scratch without facing any form of complications whatsoever.

However, one thing you should bear in mind before moving onward is that the placement of your Dot should be where it is not facing any form of obstruction (transparent or solid) within 8-9 inches of its radius.

With that said, let's teach you how to fire up your device!

Step 1

The first in your conquest should be to download the free Alexa App on your smartphone or tablet, depending on what you need.

The app is readily available in all of the leading app stores, including:

- Apple App Store
- Google Play
- Amazon App Store

Just look for the compatible one for your device and you are good to go!

Step 2

Assuming you have already set up the device in an appropriate position, you are now ready to start up the device.

- Plug the provided cable into your Dot.
- Then into your power outlet.
- Soon you will be greeted with a blue ring, which will later turn into an orange one, letting you know that the Dot is now ready to be set up.

At this moment, you will also be greeted by Alexa.

Step 3

Next, you will want to connect your Dot to your Wi-Fi network.
- For this, the first step is to go to your phone and open up your Alexa App.
- Go to Settings.
- Here, a list will pop up where you will have to find your Dot device and select the "Update Wi-Fi" option. However, if you are attempting to add a new device to your Amazon account, you will need to choose the "Set Up New Device Option" here.
- At this point, you will need to hold the Action button of your device for 5 seconds until the light yet again changes to an orange color and your cell phone automatically connects to your Dot.
- The list of all the available Wi-Fi networks will appear. Select the one you desire, enter the password and you are all set!

Step 4

Now you can actually start using your Dot. Have patience as we will soon be exploring all the features of the device. For now, to start your experimentation, do this:
- Utter the wake up word, which is by default set at "Alexa."
- Once she hears it, she will wake up and will be open to obeying your every command.
- Just for the sake of fun, why don't you try saying, "Alexa, tell me a joke!"

Chapter 2: The Primary Features of Your Echo Dot

Since you now know the most basic ins and outs of your shiny new Echo Dot, you are now finally ready to go ahead and explore whatever the device can offer! This chapter will be your first step toward unlocking the true potential and secrets of your device!

Hold your horses and your excitement right there, though. It is of the utmost importance that you understand I won't be able to immediately toss you into the more advanced concepts.

Similar to a fish who has been suddenly transported from a small aquarium to a large one, I am going to first introduce you with the key concepts of the device that should give you more confidence and make you more adaptive when trying to appreciate the features.

Now all that remains is to prepare yourself for the adventure and wake up your device by uttering the magical wake up words. Which, assuming that you didn't fiddle around with her, is "Alexa." Once you have done that, she will soon open up her metaphorical ears and prepare herself to listen to whatever you may have to say.

Let the Music Go Wild

It should be noted that first and foremost, the Dot has a built-in speaker that delivers sounds on par with other micro Bluetooth speakers, so you don't need an extra speaker! The sound quality will be lacking and might not fill a room with sound, but many reviewers agreed that it sounded better than expected.

Without a doubt then, that is the very first feature you should know about! Here, I will show you how you use your Dot to fire up the party by turning it into a fantastic juke box that is able to spur out rhymes faster than Eminem!

Given all the complex set of features Alexa is able to perform, one might naturally imagine that playing music might require the user to go through a number of different convoluted processes. However, thanks to the innovative programming on Alexa, it is much easier than one might expect!

Even at the time of this writing, streaming services compatible with the Dot are gradually increasing. One might say that the Dot has a substantial and ever-evolving ecosystem that allows it to stay "alive."

The following music streaming and radio services are currently available. Some of these are free to use, while others might require you to go for a subscription.

- Amazon Music
- Prime Music
- Spotify Premium
- Pandora
- TuneIn
- iHeartRadio
- Audible

Now, before you are able to utilize the features of any of those programs, be aware that all these service are, in reality, coming from different providers. You will need to open up/set up/link your existing account with your Dot. Doing that is pretty simple:

- Fire up your Alexa App on your cell phone.
- Press the button on the upper right side of the screen, which sports three bars.
- Here, assuming that you want to set up your Pandora account, just look for Pandora there. Click on the button and set up your account following the on-screen instructions.

Through the same procedure, you will be able to add or remove accounts according to your will. Once you are done setting up your whole musical infrastructure, the next step will only require you to know the voice commands used to control the device, which are given in the table below:

Amazon Echo Dot in 1 Hour

To do this...	Say this...
Adjust the volume	"Volume up / down." "Set volume to level [number]."
Hear details about the track currently playing	"What is this?" "Who is this?" "What song is this?" "Who is this artist?"
Halt the currently playing track	"Stop." "Pause."
To play your favorite song	"Play." "Resume."
Manipulate your sleeper timer	"Set a sleep timer for [x] minutes / hours." "Stop playing music in [x] minutes / hours." "Cancel sleep timer."
Change to either the previous or the next track	"Next." "Previous."

To do this...	Say this...
Loop the music queue**	"Loop."
Shuffle songs or tracks**	"Shuffle." "Stop shuffle."
Repeat songs or tracks**	"Repeat."
Play Prime Music	"Play [song / album / artist] from Prime Music." "Play some Prime Music." "Play a Prime Playlist." "Play [station name] from Prime." "Play [playlist name] from Prime Music."
Play Spotify Premium **(Compatible only with the Amazon Tap, Echo Dot and of course, the Echo)**	"Play [song name] from Spotify." "Play [song name] by [artist] from Spotify." "Play songs by [artist name] from Spotify." "Play music composed by [composer] from Spotify." "Play [playlist name] from Spotify."

Amazon Echo Dot in 1 Hour

To do this...	Say this...
	"Play 'Discover Weekly' playlist from Spotify." "Play [genre] from Spotify." "Play Spotify." "Spotify Connect. / Connect to Spotify." (enables Spotify Connect)
Adjust the volume	"Volume up / down." "Set volume to level [number]."
Hear details about the track currently playing	"What is this?" "Who is this?" "What song is this?" "Who is this artist?"
Halt the currently playing track	"Stop." "Pause."
To play your favorite song	"Play." "Resume."
Manipulate your sleeper timer	"Set a sleep timer for [x] minutes / hours."

To do this...	Say this...
	"Stop playing music in [x] minutes / hours." "Cancel sleep timer."
Change to either the previous or the next track	"Next." "Previous."
Loop the music queue**	"Loop."
Shuffle songs or tracks**	"Shuffle." "Stop shuffle."
Repeat songs or tracks**	"Repeat."
Play Prime Music	"Play [song / album / artist] from Prime Music." "Play some Prime Music." "Play a Prime Playlist." "Play [station name] from Prime." "Play [playlist name] from Prime Music."

Amazon Echo Dot in 1 Hour

To do this...	Say this...
Play Spotify Premium (Compatible only with the Amazon Tap, Echo Dot and of course, the Echo)	"Play [song name] from Spotify." "Play [song name] by [artist] from Spotify." "Play songs by [artist name] from Spotify." "Play music composed by [composer] from Spotify." "Play [playlist name] from Spotify." "Play 'Discover Weekly' playlist from Spotify." "Play [genre] from Spotify." "Play Spotify." "Spotify Connect. / Connect to Spotify." (enables Spotify Connect)

Radios are back

Sometimes you might want to go back in time and listen to the radio, going old school to find out what the next hot DJ is about to play! Even though time has passed, the evolution of radio has helped it to stand the tests of time! And by evolution, I mean constantly entertaining services like TuneIn and iHeart, which are winning hearts all around the world.

Radio manipulation is no different from playing your music. Just follow the voice commands listed below to please your heart's desire:

Play a custom station (iHeartRadio, Pandora, and Prime Stations)	"Play my [artist / genre] station on [Pandora / iHeartRadio / Prime Music]."
Play a radio station (TuneIn and iHeartRadio)	"Play [station frequency]." "Play the station [station call sign]." "Play the station [name of the station]."
Playing a podcast program	"Play the podcast [podcast name]." "Play the program [program name]."

Audiobooks Never Felt More Real

The built-in voice synthesizer module combined with the extremely sophisticated and advanced program behind Alexa allows your Dot to turn into the perfect reading companion whenever you might need one, all without the requirement of an irritating human being who might just constantly keep nagging you!

At the time of my writing, Alexa had full support to read out loud books not only from the Kindle Library and Audible, but also from your personal collection of PDFs or similar files you might have saved to your hard drive or cloud network.

The Dot is also able to read out potent information from Wikipedia pages from when you might require a quick jumpstart to your brain.

The following voice commands will allow you to control Alexa's reading capabilities:

Amazon Echo Dot in 1 Hour

To do this...	Say this...
Listen to an audiobook	"Read [title]."
	"Play the book, [title]."
	"Play the audiobook, [title]."
	"Play [title] from Audible."
Pause the current audiobook	"Pause."
Start-up and continue your most recent book	"Resume my book."
Forward or rewind your book by about 30 seconds	"Go back."
	"Go forward."
Go to the next or previous chapter	"Next chapter."
	"Previous chapter."
Go to a specific chapter	"Go to chapter [#]."
Restart a chapter	"Restart."
Manipulate your timer	"Set a sleep timer for [x] minutes / hours."
	"Stop reading the book in [x] minutes / hours."
	"Cancel sleep timer."

Joel Goodwin

Traffic Updates Right at Your Finger tips

Perhaps one of the biggest drawbacks of our modern world is that, with the increasing number of automobiles on the road, it is becoming increasingly difficult to go anywhere without facing an hour-long traffic jam.

With the power of your Dot, you will no longer need to face this hassle, as the Dot comes packed with the power to give you all the traffic information you might need to make sure you get a smooth sail to the nearest McDonalds!

Imagine your Dot and Alexa as being a total overseer of everything out there!

You will need to follow a few tiny steps to set up your Dot and bestow upon it the capability of giving you your traffic information.

- Take out your cell phone and fire up the Alexa App, and then look for the "Traffic" option.
- In that option, you will be greeted with two blank spaces. One will have "To" and the other will have "For." Here you are going to want to enter both the locations so Alexa can figure out the route she will need to examine.
- Save the changes once you are happy, and exit.

Once you are done with that, the following commands are going to help you get the traffic updates of your dream!

To do this...	Say this...
Ask for a traffic update	"How is traffic?" "What's my commute?" "What's traffic like right now?"

All the Weather News

No one wants to go out on a sunny day and return home completely drenched in a storm, right? The Echo Dot has the capability of accurately predicting the weather conditions of your locale by utilizing the services of AccuWeather.

If you have already set up your location, you are good to go. If not, follow the previous tutorial on this list. Once the location is set, you will be able to use the following commands to assess the weather conditions:

To achieve this	Say this
Get to know the current weather conditions	"What's the weather?"
Get to know the weather of a specifically targeted day	"What's the weather for this weekend?" "What's the weather for this week?" "What's the weather for [day]?"
Get to know the weather of another city of your choice	"What's the weather in [city, state or city, country]?"
Know about future weather conditions	"Will it [rain / snow] tomorrow?" "Will it be windy tomorrow?"

Joel Goodwin

Sports Update

Having a busy lifestyle sometimes forces us to sacrifice our favorite pastimes. While it might be a little difficult to sit in front of the tele to enjoy your favorite sports team go for the gold, that's no reason to get sad, especially not with Alexa here! Using Alexa, you will be able to get the latest updates from your favorite sports team regardless of your current conditions. Even if you are having a good time sitting in your washroom, just shout out for the news on your team!

Features are always increasing, but while writing this, Alexa had the support for the following leagues:
- EPL - English Premier League
- MLB - Major League Baseball
- MLS Major League Soccer
- NBA - National Basketball Association
- NCAA men's basketball - National Collegiate Athletic Association
- NCAA FBS football - National Collegiate Athletic Association: Football Bowl Subdivision
- NFL - National Football League
- NHL - National Hockey League
- WNBA - Women's National Basketball Association

To add your favorite sports team, all you need to do is:
- Open up your Alexa App
- Go to Settings > Sports Update
- Punch in the name of your sports team and choose one from the provided list
- Select the team and add it to "Sports Update"

Once you are done with all the formalities, the only voice command you will need to know is "Give me my sports update!" and you are done! Just don't shout out too loudly in excitement.

Allowing Alexa to Wake You Up

No more will you need to rely on old-fashioned cell phones or even your wife or mother to wake you up from your slumber! You can just simply hand over the task to dear Alexa. The Dot will give you the option to set up your alarms and manage them with incredible ease. Even if you are completely exhausted from a day of strenuous work, you can easily give the burden of waking you up right on time to Alexa. Without worrying about anything else, this device will allow you to have a good night's sleep and have the dream of a lifetime!

The voice commands required here are as follows:

To achieve this, either use these voice commands	Or set them up through your Alexa App
For setting up one alarm "Wake me up at [time]." "Set an alarm for [time]." "Set an alarm for [amount of time] from now."	Once you have used your voice to set up your alarm, you can easily go to your app and further configure it
Go for an alarm that will repeat itself "Set a repeating alarm for [day of week] at [time]." "Set an everyday alarm for [time]."	Assuming that you have an alarm set, go through the following steps: • Go to the navigation panel and select Timers • From there, find your device • Go to the tab named Alarms • Choose your current alarm • Then go to the repeats section and select one from the options in the list • Once done, finally save the changes

To achieve this, either use these voice commands	Or set them up through your Alexa App
Snooze the alarm Say "Snooze." (while the alarm is sounding). **Keep in mind that the alarm will keep snoozing for about 9 minutes**	Still not configurable
Check the status of your alarms "What time is my alarm set for?" "What alarms do I have for [day]?" "What repeating alarms do I have?" **Keep in mind that if you have multiple alarms set, Alexa will read out the alarms and let you choose which one you want recited**	- Go to the Timers section - From the menu, choose the tab named alarms - Choose the alarm you are looking for - Choose the option you need
Prevent an Alarm from buzzing up "Stop the alarm." (when alarm is sounding) "Cancel alarm for [time] on [day]." (turns off the alarm, but does not delete it).	- Go to the navigation panel and select Timers - From there, find your device - Go to the tab named Alarms - Find the Alarm you want to disarm and either remove it or delete it
To Obliterate an Alarm	

Amazon Echo Dot in 1 Hour

To achieve this, either use these voice commands	Or set them up through your Alexa App
Through the App	<ul><li>Go to the navigation panel and select Timers</li><li>From there, find your device</li><li>Go to the tab named Alarms</li><li>Find the Alarm you want to disarm and either remove it or delete it</li></ul>
Alter the Alarm Volume Through the app.	<ul><li>Go to the navigation panel and select Timers</li><li>From there, find your device</li><li>Go to the tab named Sounds</li><li>Gently press the volume bar and set the alarm volume</li></ul>***Keep in mind that changing the alarm volume won't alter the voice volume of your device**
Alter the Alarm Sound Through the app	**For brand new alarms**<ul><li>Go to the navigation panel and select Timers</li><li>From there, find your device</li><li>Go to the tab named Sounds</li><li>Choose your different sound and change it</li></ul>**Please note that this won't change the sound of existing alarms; instead it goes for the newer ones. For existing ones, follow the instructions below:**

To achieve this, either use these voice commands	Or set them up through your Alexa App
	For an existing alarm - Go to the navigation panel and select Timers - From there, find your device - Choose the alarm you want to alter - Choose your alarm sound - Save changes upon returning to the former alarm screen

The Short Cat Nap Burst

Who wouldn't want to hibernate for days on end? However, sometimes you might need to take a short power nap before waking up and getting back to work on that masterful project of yours!

In fact, it has been scientifically proven that short naps help give the body a full boost of energy to keep working throughout the day!

However, the problem arises from the fact that having someone around who might put up with you in order to wake you up just after 10 minutes could be tough. But you shouldn't worry about that anymore, especially now that you have Dot by your side. No longer will you need to go through the nuisance of asking someone to wake you up; instead you can simply set up a timer for your Dot to wake you up whenever you might need! Not only that, the timer can also be used to help you remember something, like perhaps taking the pot of boiling milk from the stove after 10 minutes?

With the following voice commands, you will be able to turn Alexa into your personal ticking timing assistant.

Amazon Echo Dot in 1 Hour

To achieve this effect, utter these	Alternatively, use your app
For setting up a countdown timer "Set a timer for [amount of time]." "Set the timer for [time]." **Keep in mind that it is not possible to set your timer more than 24 hours**	Just use the power of your voice to set it
For pausing up or continuing your timer Done through the app itself	• The first step is to go the navigation panel named Timers and Alarms • A drop down menu will come and from that menu choose the name of your device • Go to Tab of the timers • Choose the edit section next to the timer you want to edit and go to pause
Get the information of your countdown timer "How much time is left on my timer?" (tells you how much time is left on your next upcoming timers) **Keep in mind that if you have multiple timers then you will want to get the information through the app**	• The first step is to go the navigation panel named Timers and Alarms • A drop down menu will come and from that menu choose the name of your device • Go to Tab of the timers • Choose the timer and check out the information of your timer

To achieve this effect, utter these	Alternatively, use your app
Prevent a countdown timer from counting "Stop the timer." (when timer is sounding). "Cancel the timer for [amount of time]" (for upcoming timers). **If for some reason you have two timers that are set to expire within the same time, then go to the app and adjust it**	• The first step is to go the navigation panel named Timers and Alarms • A drop down menu will come and from that menu choose the name of your device • Go to Tab of the timers • Go ahead and click the edit button beside your preferred timer
Alter the countdown timer volume Done through the app	• The first step for this is to go ahead and choose the settings • From that list choose the device • And then go for sounds • Find the volume bar designated for the Alarm and Timer and drag it to your desired level **Keep in mind that the volume of the countdown does not affect the volume of the device**

Creating a Bridge Between Google Calendar and Alexa

We lead a busy and hectic life, and remembering everything can be difficult! All the bucket loads of different dates and meetings, birthdays and what not can be overwhelming.

Plus, some of us just have a terrible memory. There's no shame in asking someone (or something) to help you out and keep you updated.

That is precisely the reason why Google Calendar and similar apps were developed: to help you remember the things you just simply can't. If you happen to have a Dot alongside a fully packed up Google Calendar, you can now seamlessly blend the two of them together and synchronize your calendar with your Dot to allow the Dot and Alexa to remind you of all your important dates.

Follow these steps:

- Open up the Alexa App
- Go to the Settings option and then Calendar
- Select Google Calendar
- Press the Link Google Calendar Account and enter your credentials

Now that you are set up, follow the voice commands below to get all the day-to-day information of your life!

To Achieve This	Utter This
Get the information of your up-coming event	"When is my next event?" "What's on my calendar?"

To Achieve This	Utter This
Get to know the exact time of your upcoming event	"What's on my calendar tomorrow at [time]?" "What's on my calendar on [day]?"
Follow these steps to add in a new event to your calendar	"Add an event to my calendar." "Add [event] to my calendar for [day] at [time]."

With this you are finally done with all of the most basic and must-know features of your Dot! Now that you have finally justified your Dot purchase, the time has come to learn the more advanced concepts of your Dot!

Chapter 3: The Advanced Secrets of Echo Dot

Now that you are perfectly clear on the basic features of your device, we are going to look further into the more unusual and lesser-known or advanced features of Echo, which might be new to you at first glance.

Dot as Your Bluetooth Speaker

Despite having all these heavy functionalities, Echo Dot still is at its heart a simple Bluetooth speaker designed to play any music you want at your convenience. Now that we have already pointed out the essential "smart" features of your device, let us go a bit old school and teach you to use your Dot using just your phone to play music. The hassle of having no internet connection won't be a problem any longer, as will you be able to bring the party anywhere you want! This is achieved through the immensely stable Bluetooth technology that allows it to turn into a fully-fledged 360-degree surround sound speaker!

Connecting all the way through to the internet won't be any problem here at all, as this is one of those features that can seamlessly be used while you are offline. Through the highly sophisticated and state of the art Bluetooth technology for effortless connectivity, the Dot is capable of delivering astonishingly clear sound, which may not be as great as the Tap or Echo, but is still strong on its own accord.

Follow these simple steps when using your device as a speaker:
- The first step to do this is to make sure the Bluetooth on your device is turned on and is set to pairing mode. Once done, bring the device close to your Dot.
- Say "Pair" to prepare your Dot to be paired.
- Open up the Bluetooth menu of your phone and choose "Dot." Pair up with it and you are ready to roll!
- To terminate the connection, just say "Disconnect."

Joel Goodwin

Creating Your Very Own List

As surprising as it may sound, the Dot is actually capable of creating your whole shopping list or creating a list that will contain your life goals! We all love to tick off items in a checklist, right? Why not create a virtual one then?

To achieve this	Utter this…
Creating your shopping list by adding an item	"Add [item] to my Shopping List." "Put [task] on my To-do List."
Go through your list to double check it	"What's on my Shopping List?" "What's on my To-do List?"

When you own a dot, you'll never have to carry around a pen and piece of paper to write your list, and you won't need your smart phone, either! Once you have set up your Dot and a list, you can simply use the power of your voice to let Alexa know whatever you want her to do with the list. You can add, remove, check or uncheck items. What's more is you can even print out your lists in a jiffy through the Alexa App and your computer.

Below are the voice commands you are going to need to know to manipulate Alexa and your Dot into creating your cherished list.

To perform this action	Perform this action
Print out your created list	Using your computer and Alexa App • Simply go to your navigation menu and choose the panel title "Shopping & To-Do List"

Amazon Echo Dot in 1 Hour

To perform this action	Perform this action
	- Select the list you are looking for - Go to print
Open up an existing list	**Using the Alexa app** - Simply go to your navigation menu and choose the panel titled "Shopping & To-do list" - Select the wish you are looking for to open it up **Keep in mind that having an internet connection is not essential to view your list.**
Add an additional item to your created list	**Using the Alexa App** - Just go your preferred list - Enter the name of your task/item - Click on the plus symbol
Alter an item from your existing list	**Alexa app** - Go to your list and head over to the item you want to modify - Once found, click edit
Obliterate an item from your existing list	**Using the Alexa App** - Press the arrow that is placed next to your desired item (a downward facing one) - From the menu, choose delete to remove that item.

To perform this action	Perform this action
Select an item and deem it complete	**Using the Alexa App** This is straightforward, as it only requires you to choose the checkbox and click on it to toss in under complete

A Bliss for Big Families

Alexa is great when it comes to taking care of different family members. Whether it be a large family or a small one, the taste and musical preferences of each and every person might vary. While you might be a fan of deathcore, your sister might be a fan of soft classical music. Therefore, cramming all of your preferences and music stations in just a single Alexa profile might be somewhat of a nuisance. Keeping that in mind, Alex has been designed to blissfully manage different profiles for each house member through the Alexa App. Juggling between these multiple profiles is without any form of complication.

For adding a new profile

The following steps will simply allow you to add in a new profile to your household.

- From the navigation panel just go to Setting
- Then go to Account and choose the household Profile option
- Click there and you will be greeted with a number of instructions explaining how to add a second person to your Household
- Follow them properly and you are done

Removing a profile from your household

Similar to adding the profiles, you can easily remove any profile from your household as well.

- From the navigation panel, go to Settings

- Then go to Account and choose the Household Profile option
- Choose the person you want to obliterate
- Press the Remove from Household button and the profile of that person will go away

It should be noted though that once you have removed a profile, you won't be able to add it back for the next 180 days. So, this is definitely something you should be aware of.

If you ever accidentally remove your own profile or a profile of someone you were not supposed to, your only solution would be to call customer care support and follow their instructions to solve your problem.

Alter between profiles on the fly

Once you have multiple accounts in your Dot, now you might need to know how to juggle between the different profiles set in your device. Do the following:
- Utter the words "Switch Account"
- Alternatively, while you are going through the content library in your App, just click on the drop down menu where you will get the options to toggle between libraries

Changing the Name of Your Dot

Some of you might imagine that the name "Alexa" isn't sweet enough, or your device might feel neglected when you call her Alexa. Or the more important issue is that your girlfriend or wife is getting jealous and scolding you from time to time when you utter the name of another girl.

There's a simple solution to that problem. Just change the "wake up" call word of your device and save your love life from going down into utter anarchy.
- The first thing to do here is to turn your cell phone up and fire up the Alexa App. Then, go to your settings and choose the name of your Dot.
- Once you have found your device, go for the option called "Wake Word" and click on that. It will give you the option and instruction to set up your new wake word.

It should be noted though that upon changing the wake up word, Alexa/Dot will remain dormant for about 30 minutes. It takes this time to properly re-adjust herself and get habituated to the new call sign.

During this period, however, the device will remain inoperable, so don't be afraid! This is normal.

Delete All of Your Recordings

Any device that has the capability to keep "listening" to whatever we are saying is bound to face some controversies, which was also the case for the Dot. Naturally, when the news of Amazon storing all of our voice recordings came out into the wild, people began to suspect the device for being just another CIA mind trick to listen to what we were saying!

This was definitely not the case. The true story here is that these recordings are not shared at all and are stored under a private folder based on the user's account. The Dot and Alexa uses these recordings to evolve itself further by assessing the user's voice pattern and other preferences.

Even still, for those of you out there who might feel insecure, it is easy to remove these recordings just by following these steps. Keep in mind that removal of these might cause a decline in the quality of your experience.

Just go through the following steps:
- Go to your Alexa App and open Settings
- Then go to History
- You will be greeted with all the requests and commands you have issued since the birth of the device.
- Click on the recording you want to delete, and it will be done!
- Or if you want to clean everything up in one go, visit **www.amazon.com/myx** and sign in there.
- Select your device and go to the "Manage Voice Recording" options where you will have the button to remove everything at once.

Some Fun Games to Jump Into!

Some might think I'm being sarcastic here with the title, but I am actually not! The Dot is not just meant to be a serious powerhouse of functionalities. It can sometimes just be a fun and helpful companion for when your younger siblings might be around.

- **Coin Flip:** Since the very early and ancient days, the process of flipping a coin has been used to settle countless bets and wagers and destroy or mend friendship alike. If you are in need of a coin and don't have one, then just go ahead and ask, "Alexa, Heads or Tails?" and the device will respond promptly!
- **Bingo:** Bingo is nothing short of a classical family board game that stimulates the friendly nerves as well as the funny bones. Make sure you have downloaded a proper Bingo Card from the internet to set yourself up. Once you are done with that, just say, "Alexa, play Bingo!" and you are in for the thrill of a lifetime.
- **Simon Says:** Simon Says is yet another classical game that has been re-introduced into the Dot for an extra added "oomph" factor. To utilize this program, just say "Simon Says" before saying anything to Alexa, and she will repeat it back to you. Creating the opportunity for you to go for a nice Simon Says Sequence.
- **Simple Geography Quiz:** Geography tests are boring, we all know that! But when you are playing the game with your friends and Alexa, things are bound to get spiced up, which will lead to some interesting and memorable times. To fire up a session of 20 interesting geography questions, all you need to do is say, "Alexa, start animal game/capital quiz," and Alexa will start throwing questions at you like there is no tomorrow.

Joel Goodwin

Finding the Location of Your Package

Amazon is really cranking up your shopping game and thrusting themselves into the future with the announcement of their new Amazon Go shopping mechanism. It seems like when it comes to customer satisfaction, they are not leaving any stones unturned!

Following that motto, they did add a nifty feature to Alexa and Dot which would seamlessly allow you to get constant updates on any package you might have ordered from Amazon.

This software is still in its early stages and requires some work, but from what I have seen so far, it works pretty adequately, making the wait for the next big video game even more restless!

- Add the link to your ordered package in your Alexa App.
- Once you are done, whenever you want to know about the whereabouts of your package. Just say, "Alexa, where's my stuff?" and she will promptly reply by letting you know where in the universe your package is residing.

Your very own kitchen assistant in the form of Echo Dot

Using your Alexa, you will most definitely be able to read out loud the next groundbreaking book of 100 Paleo recipes, but that doesn't mean you won't need some help figuring out the proper quantities while cooking! The Dot can easily be converted into a pretty handy kitchen assistant who will be able to help you around your kitchen Like with simple calculations such as the exact amount of salt or flour required for the next culinary masterpiece.

Example voice commands include:
- Alexa, how many cups make a quart?
- Alexa, how many teaspoons are in a stick of margarine?
- Alexa, set the timer to 10 minutes
- Alexa, convert 15 milliliters to liters

And similar ones.

Using Echo Dot as your personal calculator

Throughout my test session, I gave Alexa two problems to solve:
"Alexa, what is two plus two?"
"Alexa, divide 111.11 by 66.66"
Needless to say, all of the results were mind numbingly correct. The interesting thing to point out here is that Alexa is actually able to not only deal with integers, but also floating numbers (numbers with decimal points), which is actually really interesting, making this device even more versatile!

Regardless of the problem, Dot could tackle it with ease. One thing that surprised me about the Dot was the fact that it was capable of dealing with floating numbers (decimal numbers), which I wasn't quite expecting. This just further helps to make the device even more versatile in nature!

The Power of Dot in the Form of Skills

Even though I have touched on some really interesting features for the Dot, you haven't seen anything yet! Now I will tell you about what makes the Dot truly special.

Similar to the Google Play Store, Amazon has created a store that is designed specifically for Alexa, in the hopes of extending its reach and adding more functionalities in the future. Basically, anything you can dream of, if it can be represented in an audio form, chances are Alexa will be able to do that! These "Skills," as they call them, are what makes Alexa truly special.

At the time of this writing, the Skills store had more than 1,500 skills ready to be downloaded, but the list keeps growing every day!

Given the amount of great Skills available on the store, it won't be possible to cover all them in this book, but rest assured I have taken the time to give you some of the best and more interesting skills out there.

First, you must activate your skills:
- Open up your Alexa App
- Go to Settings
- Using the search bar, enter your desired skill or use the "Categories" section to browse through the available skills

- Once you find a skill you are happy with, click enable

As for the list, here we commence:

- **Fitbit:** This is undoubtedly the best possible activity tracker app present in the app store. The app allows you to blissfully keep track of all the various information, such as steps taken, your exercise goal or even sleep tracking to make sure you lead a happy and normal life!
- **The 7-Minute Workout:** If you are into fitness, you should definitely check out this app, as it allows you to keep track of a specific exercise routine, all of which can be followed within just seven minutes or so.
- **Stock Exchange:** This is one app that administers to the budding businessmen out there. With this you will be provided with a plethora of inputs to be inserted, and you will be able to obtain a summary of all the chosen stocks that help you administer them into a larger portfolio.
- **The Bartender:** This is a pretty unique app. Even though the app store is full of ups and downs, this is one of the better ones, undoubtedly! The brilliance of this app allows you to simply tell Alexa to narrate the recipes of all, if not most, of the famous bar drinks out there.
- **Yo Mamma Jokes:** Keep in mind that this app is not suitable for all ages, but if you're an adult, this will definitely be of interest to you. The app does just as the name implies: it shoots out a large number of "Yo Mamma" jokes whenever you want to hear them! Perfect for a good laugh amongst friends and co-workers.
- **4A Fart:** While the apps in Echo Dot can seem a bit serious, this one is all for the jokes! The app, as its name implies, goes all out to produce horrendous fart noises at your command.

- **The Magic Door:** This is an extremely cool app that will allow you to dive deep into the realm of magic through a grossly enriched narrative. The main purpose of this app is to help your child open their imaginative potential by going through different storylines, all of which are extremely interesting and even come with immersive sound effects and multiple choices to influence the ending of the tale.
- **Akinator:** This is one of the most intriguing games present in the Skill Store. This skill allows you to play a nice game of 20 questions with Alexa, where Alexa has to guess the character you are thinking about!
- **Amazing Word Master Game:** This app allows you to enhance your vocabulary through a game. This app will simply throw random words at you and allow you to give the meaning of them and train you while you are at it.
- **Pick-up Lines:** Are you so hopeless and desperate in your life that you always have to reach out to your friend for life advice on how to impress the girl you love? Well, you won't have to embarrass yourself ever again, as Echo Dot has got you covered. This app will simply keep throwing different pick-up lines at you until you turn into a complete pro.
- **Capital One:** An app sponsored by the Capital One Bank, this program actually allows you to control and get all the credentials of your Capital One bank account. With this, you will be able to check for information on transactions, as well as pay credit card bills with your voice.
- **Campbell's Kitchen:** An app from one of the most well-known and trusted food companies out there! The app produces a wide variety of recipes that range from soup, meat or even fish. If you are a budding/aspiring chef, you don't want to miss out on this.

- **TV Shows:** The function of this app is pretty much self-explanatory. If you are a TV hog, then this is the ultimate skill for you! The app will provide you with all of the accurate information regarding your favorite TV series and fire at you all of the necessary information, such as the timing of the next episode.
- **Automatic:** With Google leading the industry when it comes to creating automated vehicles, this app is Dot's first step into this world. The Automatic skill allows you to sync with your vehicle pretty nicely and provide you with different stats, such as distance driven, location and gas.
- **Uber:** The world-famous taxi calling app is also available in your Echo Dot as well! As you may know, Uber helps to catch a ride, but now you will be able to do it by just uttering a few words. Hire a ride, check your ride status and even check information on the ETA of your hired ride.
- **1-800-Flowers:** Sometimes you might want to surprise your fiancée, your mother or your wife with a beautiful bunch of roses, but our busy lifestyle can prevent us from buying them in time. Using this app, you will be able to make your arrangements easily. Just poke your Alexa, and this skill will help you choose the flowers, your destination, and the date of delivery.
- **Domino's Pizza:** Definitely one of the leading Italian restaurants out there! With this app, you can simply order your cheesy and yummy pizzas in just a moment's notice. The device will even alert you whenever your pizza has arrived.

- **The Wayne Investigation:** Similar to the Magical Door app, this skill also focuses on creating a large and engrossing story through simple narrative techniques. While the Magical Door was for children, this one is for a matured audience. Using this skill, you will able to play out this game, but here's the catch: you play as the world-famous detective Batman to solve an interesting case. This one also has multiple endings, and sound makes the experience much more immersive. It should be noted that this is amongst the top rated games in Skills.
- **Jeopardy J6:** Any game show enthusiast would easily recognize the name of this game in a jiffy. This app is based on that very show, and no other app does trivia better than this one! Not only will you have fun while playing this game, but you will also get a boost in your general knowledge as well.

Chapter 4: Automating Your Home

This is where the "smart" part of Echo Dot comes into play. After reading this chapter, you will know exactly how much potential your device has and how to utilize it.

Besides all the gorgeous services and skills available in Dot, the device can also be used to completely allow you to manipulate the automatic electronics and gadgets in your house in a jiffy. Just have a look at the Jetsons because that is the future you are headed toward with this tech. And the best part? All of your devices can be controlled by only your voice! How cool is that?

Use Dot as the centerpiece of your automated home. You will be able to control the lighting in your house, the room temperature, security panels, and power functions of specific electrical components as well. Interested in knowing how all of these work? I'll tell you!

The secret behind everything

The reason why Dot can break down all the barriers between human and electronic interaction is nothing more than the advanced integrated service known as the IFTTT. While the name may sound just a little bit weird, this is actually the software powering the interaction system of the device.

The service is essentially handcrafted from a number of different formulas, which helps Dot to understand human interaction and get to know how it should communicate with another device or service. For example, the IFTTT allows Dot to connect with a Belkin WeMO switch in order to let you manipulate the power level of your lamps.

The list of compatible devices

Before actually moving onto teaching you how to add the various devices, I thought it might be helpful to give you a quick list of most of the devices compatible with Dot up until the latest update. This should help you start up your journey and take baby steps toward home automation.

Smart Home Hubs

- Samsung SmarThings Hub
- Wink Hub
- Insteon Hub
- Alarm.com Hub
- Vivint Hub
- Nexia Home Intelligence Bridge
- Universal Devices ISY Hubs
- HomeSeer Home Controller
- Simple Control Simple Hub
- Almond Smart Home WiFi Routers

Lighting

- Philips Hue White Starter Kit
- Philips Hue White and Color Ambiance Starter Kit
- Philips Hue Go
- Philips Friends Of Hue Lighting Bloom
- Philips Friends Of Hue Lightstrip
- LIFX Color 1000 A19 Smart Bulb
- LIFX Color 1000 BR30 Smart Bulb
- LIFX Color 800 A19 Smart Bulb
- LIFX Color 900 BR30 Smart Bulb
- Cree Connected LED
- GE Link Bulb
- OsramLightify Smart Bulb
- TCP Connected Smart Bulbs

Switches, Dimmers and Outlets

- Belkin WeMo Light Switch
- Belkin WeMO Switch
- Belkin Wemo Insight Switch
- iHome Smart Plug

Amazon Echo Dot in 1 Hour

- Samsung SmartThings Outlet
- TP-Link HS100 Smart Plug
- TP-Link HS110 Smart Plug With Energy Monitoring
- D-Link WiFi Smart Plugs
- Insteon Switches, Dimmers and Outlets
- GE Z – Wave Switches, Dimmers and Outlets
- Leviton Switches, Dimmers, and Outlets
- Lutron Caseta Wireless Switches, Dimmers and Remotes
- Enerwave Switches, Dimmers and Outlets
- Evolve Switches, Dimmers and Outlets

Thermostats
- Nest Learning Thermostat
- Honeywell Lyric Thermostat
- Honeywell Total Connect Comfort Thermostats
- Ecobee3 Smarter Wifi Thermostat
- SensiWiFi Programmable Thermostat
- Haiku Home Ceiling Fans
- Keen Home Smart Vents

With that knowledge, it's time to finally learn how to actually connect your Dot with those devices!

Joel Goodwin

- Open up your Alexa app and click on the three bars in the top-right corner.

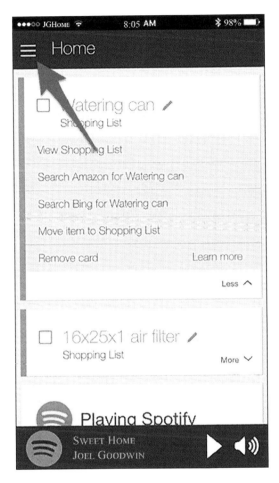

Amazon Echo Dot in 1 Hour

- Next, a menu will appear where you can click on Smart Homes.

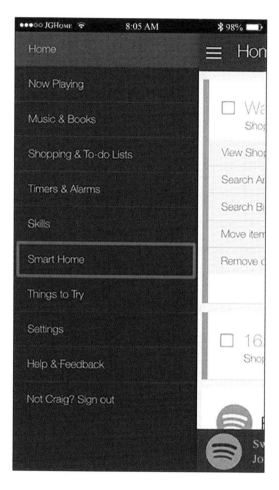

- Once there, look for the option called "Discover Devices", and click on that.

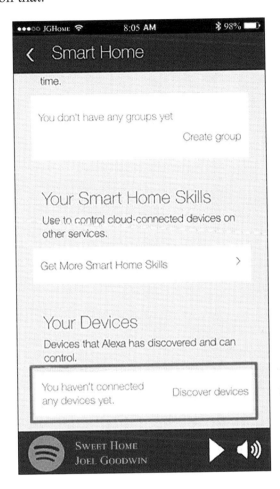

Amazon Echo Dot in 1 Hour

- Follow the menu where you will see a list of the discovered automated devices available for Dot to connect with. Just choose the device you need, and you are done!

An alternative way to add your devices

If the previous steps mentioned are giving you a hard time, it means you will need to download a new skill to get your job done!
- Go to the Smart Home screen again and click on the "Get More Smart Home Skills" button.
- Inside the Search Skills Box, type in the brand of your smart device you are trying to pair up. In this case, we are going for SmartThings and ecobee.
- Once you have found them, enable them and follow the instructions to insert the required credentials.

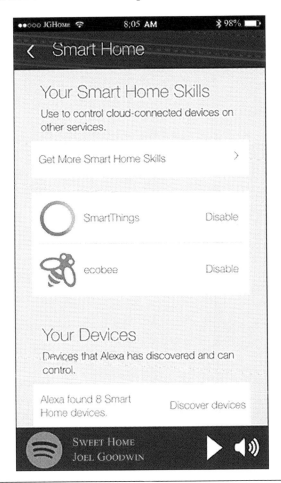

- You should now be able to look for your devices.

The Final Step

- Once everything is done, you should now be able to control everything around you. For your added convenience, it's better to add everything in a single room to a specific group.
- For example, here we have gathered all the lights in our bedroom into a group called "Bedroom." How will this help? You will see in a bit.

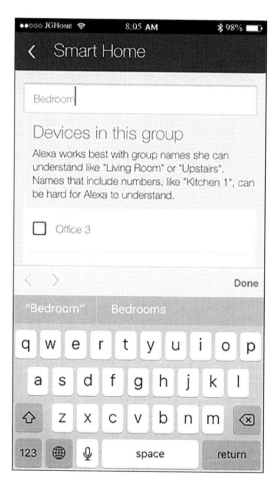

Now you should be ready to use your automated devices! Just follow the voice commands below to manipulate everything:

"Alexa, turn [groupname] [on/off]."

"Alexa, turn [on/off] [groupname]."

"Alexa, start [groupname]."

"Alexa, set [groupname] to [X%]."

Always keep in mind that you can go for the different names of your device as well to interact with them, but the group name will just allow you to make things much more accessible to you.

Chapter 5: A Selection of Troubleshooting Tips and FAQs

No device in this world is perfect! That is even more accurate when it comes to electronic devices and gadgets. Every now and then, they tend to go sour and malfunction just a slight bit. But that is exactly why updates and support teams are here! And that is even superbly done when it comes to the Dot.

After its release, Amazon has always kept its ante up in order to solve the various problems of the Dot through software updates.

Keep in mind that even after solvable issues, sometimes things might go sour and you may face some otherworld technical issues. The following chapter focuses on some of the problems you may face and teaches us how to troubleshoot them as well.

Lost Wi-Fi connection

This is one of the more common problems faced when opening up your new Dot. Due to various troubles, the Dot may not be able to get a proper connection to your Wi-Fi or might just keep getting constantly disconnected. This can easily be noticed by looking at the ring on top of your Dot. If it is showing an unblinking orange color, it means something is wrong while a solid white color means it is properly connected. If you happen to face something like this, go through the following steps and it should solve your problem:

- The first step here is to gently disconnect your device from the power cord and let it stay still for 3 seconds and then connect the cable to the device again.
- After that, making sure your device is properly linked to your Amazon account, use your Alexa App to navigate to the section where you will be asked to manage the connected devices. Look for your Dot.
- Select the Unregister option from there and finally register and start the whole process from scratch to re-new the connectivity of your device.

Unable to Find Connected Home Devices

Aside from facing normal connectivity problems, sometimes you might find it difficult for the smart devices to be found by the dot. In such situations, just go through the following methods

- The first step you should check is to make sure your smart device and Dot are on the network.
- Once done, just go ahead and update your Wi-Fi network and you should be able to find the smart devices

Another route you might take is the following, which is a more advanced one, so to say:

- Pronounce the words "Discover my devices" while making sure your devices are set in the same band of Wi-Fi frequency as your Echo Dot. Keep in mind that your Dot releases a frequency of 2.4 GHz as opposed to 5, so your devices should also be on 2.4 GHz
- Make sure to do that by changing your network router to 2.4 GHz and enabling SSDP or UPnp
- This should solve your problem
- Restart your Dot and the smart devices as well

Hard-resetting Your Dot

Sometimes you might face a situation where you may come upon trouble you are unable to solve easily. Or maybe you just altered the settings of your device, which is now bringing unforgiving and unfathomable results! In such cases, what you might want to do is follow the steps below to give your device a hard reset:

Dot 1st Gen

- Turn over your device and find the hole under the Dot. Take a pin and push it inside the hole.
- Hold it there for about 5 seconds, or until you hear a nice clicking sound.

- Once the device has been reset, the light ring on top of your Dot will light up in an orange color and stay like that for about a minute.
- After this point, your device will get back to its original factory settings where now you will be able to set it up once again from the get-go.

Dot 2nd Gen

- Find the **Microphone off** and **Volume down** buttons.
- Press and hold both buttons at the same time until the light ring on top turns orange and stay like that for about 20 seconds. Then light ring turns blue.
- Wait for the light ring to turn off and on again. The light ring then turns orange again, and your Dot enters setup mode.
- After this point, your device will get back to its original factory settings where now you will be able to set it up once again from the get-go.

Dot not being able to hear properly

Sometimes it might happen that the Dot is not able to register your voice properly, and as such is not able to follow the commands you are throwing at it. It's really easy to tackle this problem, though. Just go through the following instructions:

- Restart your speaker
- Keep in mind that the device should be placed in a scenario where it is 8 inches away from any form of obstruction from either side
- Keep it away from anything that produces ambient noise, such as an air conditioning unit
- Even after going through the above procedures, if you are still facing trouble trying to make your Dot understand, you can go ahead and download a training skill known as "Voice Training," which will further help to update the voice

recognition software by opening up its mind and helping it to understand 25 of the most common phrases used while using a Dot.

F.A.Q Section

This section is going to cover some of the common questions asked by new users of the Dot. Keep in mind that I have already discussed most of these in previous chapters, but this will still help you to pinpoint an information you might be looking for.

What are the most basic steps of using Echo Dot?

If you want to simply ignore all the complicated stuff and jump in and experiment with your device, then the only thing essential to remember is the default wake up word, which is "Alexa" followed by your desired command. The wake word helps the Dot to understand that you are referring to it. Once she hears it, the Dot will prepare itself to listen to whatever you may have to tell her and perform an action as specified.

Is it possible to improve the voice recognition services of Alexa?

This is one of the more common questions amongst people who tend to become highly curious when trying to assess the evolving voice recognition capabilities of Alexa.

The Dot already improves upon the other devices when it comes to voice recognition by having a handy speech recognition processor built in.

Keeping that in mind, the answer to that question is actually yes. As you keep using your Dot and Alexa, she will keep recorded versions of the commands you have issued. Before you jump off your saddle, these recordings are not shared in any way and are stored in your own Amazon account whereupon you will be able to delete them later.

This is not recommended, however, as Dot uses these recordings to train herself better and understand your voice pattern and preferences in order to register your voice better, while trying to come up with suggestions depending on your taste if required.

Is there any record/history of commands given to Alexa?

As with any device that takes voice input, controversies around the Echo line of products also rose claiming that this just might be another device by the CIA to listen to our conversations secretly.

But that is not the case. While the Dot does keep a record of all the commands issued, you are completely in charge of what you want to do with them.

In fact, it is actually possible to go through every single one of the recorded voices individually. Just tap on them to get more instructions.

Is it possible to delete the saved recordings?

Yes. In fact, it is possible to go through every single one of the recorded voices individually and delete whichever you want. To perform this action, you are just going to need to go to the history section of your Alexa App.

In that section, you will find individual groups containing the requests and questions that you made to the Dot.

This differentiation will allow you to easily find what you are looking for and promptly delete anything specific.

Keep in mind that unless you are facing some security issues, you should not delete previous data, as it may drastically decrease your experience with the Dot, as Dot processes this recorded data to further upgrade its voice recognition capabilities and enhance your experience.

What are Amazon Skills?

This is a common question that sometimes floats around the minds of many people. They often ask, "Will the voice recognition capabilities of Alexa increase overtime?"

The answer to that question is yes.

Just like the Play Store for Google Android phones, Amazon Skills provides a fantastically fledged out store, which allows you to bring out the additional capabilities of your Dot and enhance its generic features.

These "skills" are provided by both First Party and Third Party developers, and really help to push the Dot to its limit by opening up a whole new world of ever-growing possibilities. Want to go out on an interactive audio adventure? There's a skill for that! Want your Dot to take care of your health and help you trim down your fat? There's a skill for that as well!

How Echo Dot processes the shopping feature?

It is actually possible to order a number of Prime products from the Amazon Prime section of the shop as long as you are a Prime Member. It should be noted that you are not only limited to Prime Products, as you are open to purchasing audiobooks, music albums or individual tracks.

The default payment system for the Dot will always ask you for an audible conformation code and will allow you to check all of the information before checking out on your Alexa App. Once you have confirmed it, Alexa will place the order and you will only have to wait until your package/purchase arrives!

Is it possible to turn off the purchasing feature?

The versatility of Alexa and Dot allows the device to be used as an awesome companion for when it comes to shopping! Through the Dot, you will be able to purchase various items from eBay or Amazon using just your voice. But here lies the risk of someone accidentally ordering a product you might not have authorized!

As a security measure for this, it is possible to set up a security code mechanism.

Just go to your Voice Purchasing settings from your App first, and turn that feature off.

In that same place, you will find the feature for inquiring your specific security code. And from that moment forth, the Dot won't be able to order any products without knowing the specified security code.

Conclusion

Now that we are done with everything, it is time to end our journey here!

Given that you have properly gone through all the instructions I set in this book, you should have a strong grasp of both the basic and advanced functionalities of the fabulous Echo Dot device! Be warned, though: this book is nothing but the tip of the iceberg, and there are actually thousands more functionalities for you to explore for yourself!

With more than 1,500 skills out there, and with even more coming each and every day, there is no limit to what an Echo Dot device can do. As long as a job can be done using only audio, rest assured that Echo Dot can do it.

From this moment on, I would like you to experiment with new possibilities whenever you can. Go ahead and turn your house into an automatic home in no time, and watch as your dreams of the future come to reality.

I would like to thank you for purchasing this book, and I do hope it was an informative and fun read for you.

Good day and God bless!

Made in the USA
San Bernardino, CA
18 October 2017